MILITARY CYBERSPACE: FROM EVOLUTION TO REVOLUTION

The evolution of communications networks into the cyber warfighting domain presents challenges across the doctrine, organization, training, materiel, personnel and facilities (DOTMLPF) spectrum. The Department of Defense (DOD) created new joint and service headquarters to address these challenges and to address the cyberspace risks to national security. While these headquarters have been able to make some progress in addressing cyber challenges, the current DOD approach to cyber depends on antiquated doctrinal concepts, mission command constructs and indefensible network architectures.

No single organization in the DOD has the responsibility to build, operate and defend cyber networks. Each Service runs its own network with network operations and security centers (NOSCs) spread across the globe. These networks and NOSCs are hidden under various layers of command below the Geographic Combatant Command (GCC) and are not readily responsive to the warfighter's requirements. The current cyber force organization also makes it extremely difficult for U.S. Cyber Command (USCYBERCOM) to execute their mission of directing the security, operation and defense of the DOD global information grid (GIG). The DOD must pursue an enterprise approach to network management in the cyberspace domain to support the GCCs and enable USCYBERCOM to accomplish its mission.

This paper begins with an overview of national interests and strategy and briefly describes the security challenges in cyberspace. It then analyzes the emergence of USCYBERCOM and service cyber commands and assesses how well the new structure addresses the challenges and fulfills GCC mission requirements. The paper also evaluates the effect of current network management and architecture on cyber operations and concludes with recommended mission command, interoperability and security improvements. [1]

National Interests, Strategy and Security Challenges

The United States' ability to achieve its national interests increasingly relies on cyberspace operations, security and interoperability. This fact is highlighted by the flurry of national cyber strategy documents published in the last two years.[2] To facilitate clear lines of authority, these strategy documents divide cyber responsibility into four areas: national, military, critical infrastructure, and non-critical infrastructure. The President has overall responsibility for establishing the national cyber policy and directing crisis response. The DOD is responsible for operating and securing the military cyber domain, Department of Homeland Security (DHS) oversees national critical infrastructure, and the Department of Commerce (DOC) manages national non-critical infrastructure. While different departments manage each sector of the national cyber infrastructure, the strategy for each sector stresses the need for secure, trustworthy and resilient networks to promote and achieve national objectives.

The *National Security Strategy* (NSS) states that securing cyberspace is a key enabler to achieving national security.[3] Similarly, the *National Military Strategy* (NMS) stresses effective Joint Force cyberspace operations and defense to accomplish the national military objectives.[4] These national military objectives support the national security objectives. Focusing on the combatant commander, the NMS states that cyberspace capabilities allow commanders to operate effectively across all of the other warfighting domains: air, land, sea and space.[5] The NMS also specifically states that,

"Joint assured access to the global commons and cyberspace constitutes a core aspect of the U.S. national security and remains an enduring mission for the Joint Force."[6]

The most recent strategy document, *Sustaining U.S. Global Leadership: Priorities for 21st Century Defense*, establishes the DOD priorities in a time of fiscal austerity and transition.[7] This document also defines "the projected security environment and the key military missions for which the [DOD] will prepare."[8] Despite outlining cuts in many defense capabilities, the Secretary of Defense stated that the DOD must continue to lead efforts to secure the global commons and defend against cyber espionage and possible cyber attacks.[9] The DOD will also deter and defeat aggression in all domains, including cyberspace.[10] Finally, the DOD will invest in advanced cyberspace capabilities and develop innovative cyber concepts of operation.[11]

Commensurate with the latest directive, the *Department of Defense Strategy for Operating in Cyberspace*, defines five strategic initiatives for seizing opportunities and minimizing risks in the defense cyberspace domain. These initiatives focus the Combatant Commands, Services and Agencies on cyber requirements that enhance "U.S. defensive readiness and national security."[12] The first initiative officially recognizes cyberspace as a warfigthing domain so that the DOD can "organize, train, and equip for cyberspace as [they] do in air, land, maritime, and space."[13] To accomplish this mission, DOD established USCYBERCOM and service cyber components. The second DOD cyberspace operating initiative is to use new defense techniques in order to protect DOD systems and networks.[14] This initiative focuses on individual training, cyber best practices, and active defense. The third, fourth and fifth initiatives are partnering with other U.S. entities, partnering with like-minded nations and international organizations, and harnessing the ingenuity of U.S. citizens and rapidly developing innovative technology.[15]

Shortly after the DOD published their strategy for operating in cyberspace, the DOD chief information officer (CIO) released the *Department of Defense (DOD) Information Technology (IT) Enterprise Strategy Roadmap.* This strategy outlines a more granular plan for increased efficiency, effectiveness, and security within the five cyberspace operating initiatives. The DOD CIO specified twenty-six initiatives that will optimize the joint environment with a recognition that the current network environment "significantly detract[s] from or completely negate[s] the ability to securely share information across the enterprise and/or ... execute effective [mission command] of DOD networks. As a result, the effectiveness, agility, and security of geographic [Combatant Command] and Combined Joint Task Force Commander's networks are significantly degraded."[16]

There are three major drivers behind the national and defense cyberspace strategies: mission effectiveness, resource efficiency and security. Each of these documents emphasizes achieving national security objectives and enabling DOD mission accomplishment. However, each document acknowledges that these objectives must be achieved efficiently and responsibly within current national fiscal challenges. The *DOD IT Enterprise Strategy Roadmap* specifically identifies $5.2 billion of future savings.[17] Finally, security is the pervasive theme of every cyberspace strategy. Each document recognizes that the "[U.S.] reliance on cyberspace stands in stark contrast to the inadequacy of [U.S.] cybersecurity."[18]

Cyber Threat

Director of National Intelligence (DNI), James Clapper, cited cyber threats as the third most important threat to the nation behind terrorism and weapons of mass destruction.[19] He stated that more than 60,000 new malicious programs and variants are discovered each day and that intellectual property theft is increasing.[20] Unidentified, nefarious actors directed recent attacks at Google, a nuclear laboratory, defense contractors, the U.S. electric grid, and defense networks.[21] During these attacks, the attackers downloaded terabytes of research and development, defense contractor, and defense sensitive but unclassified information. Due to the difficulty of determining attribution in cyberspace, the intelligence community has not been able to link these attacks to a specific actor.[22] These attacks may have installed malware that adversaries could use for future, more destructive attacks on U.S. critical infrastructure.[23]

In addition to the DNI's assessment, the *Quadrennial Defense Review* (QDR) identifies cyberspace security as a near term military operational risk.[24] Because the military is extremely dependent on the network for mission command, intelligence, logistics and weapon system operation, military networks are under constant attack.[25] Future adversaries will most likely employ cyber attacks in an attempt to minimize U.S. military and technological advantages. The DOD must take action to counter the increasing "volume and virulence of attacks" and defend against future attacks.[26]

Cyber Organizations

In response to this increasingly critical and volatile domain, DOD created USCYBERCOM as a sub-unified command of United States Strategic Command (USSTRATCOM). USCYBERCOM has responsibility for managing cyberspace risk, assuring network and information integrity and availability, maintaining a common operational picture (COP) and developing integrated capabilities in coordination with the Combatant Commands and other DOD organizations. USCYBERCOM also synchronizes and coordinates the service cyber components.[27]

The service cyber components direct cyberspace operations for their respective Services. These operations include building, operating and maintaining the communications network, managing security and performing intelligence functions. Funding for these operations comes directly through Service channels. Also, the Services determine the organizational constructs required to support cyber operations. While each of the Services has a single, centralized cyber component, the Army has regional cyber centers, formerly known as NOSCs, which manage day-to-day operations in each GCC region. The Air Force and the Navy, on the other hand, have chosen to centralize day-to-day cyber operations under operations centers located in the United States.[28]

In accordance with Joint Publication (JP) 6-0, *Joint Communications System*, the combatant commands create global network operations control centers (GNCC) and theater network operations control centers (TNCC) to manage their communications networks. Similarly, the joint force commander creates a joint network operations control center (JNCC).[29] These network control centers really do not "control" anything. They collect and analyze information from the service cyber centers and create reports for the combatant commander and issue direction to the service components.

Because cyber is an emerging domain and current doctrine does not address cyber organizational constructs, each combatant command is experimenting with how best to organize their intelligence (J2), information operations (J3), communications (J6) and other staff elements to support cyber.[30] As an example, U.S. Central Command (USCENTCOM) created a Task Force 236 as part of the combatant command staff to tackle day-to-day cyber challenges in their area of operations (AOR). U.S. European Command (USEUCOM) created a Joint Force Cyber Component Command (JFCCC) during their most recent Austere Challenge exercise. The deputy J3 was the JFCCC commander and the J2 and J6 served as deputies. The JFCCC was on par with the Joint Force Land Component Command (JFLCC) and Joint Force Air Component Command (JFACC).

While the combatant commands seek ways to direct cyberspace and cyber forces, the combatant commander has very little control over cyberspace. The Services have direct control over network operations, defense and operators in the combatant command AOR. Add USCYBERCOM into the equation with their mandate to "synchronize operations and leverage current and emerging technological capabilities to provide integrated effects to strategic, operational, and tactical commanders," and it is readily apparent that, despite their best efforts, the combatant commanders have limited ability to affect cyberspace.[31] The combatant commander's inability to direct cyberspace operations inhibits the synergistic effects of combining conventional and cyber operations.

Current cyber organizational structures present even more challenges for the deployed joint force commander. The combatant commander and staff have the advantage of working with the service component cyber centers under mostly steady-state, day-to-day operations. The joint force commander, however, has to deal with operational units and a very dynamic network environment. In this environment, the

7

JNCC synchronizes tactical to strategic communications through functional and service cyber centers.[32] These functional and service cyber centers build, operate and maintain the network and cyber systems for their parent organizations and report communications and cyber status to the JNCC. The JNCC fuses the multiple inputs into one COP for the joint force commander. Direct control of cyberspace and forces, however, remains with the functional and service components.

A simple operational example will help illuminate the challenge of current cyber organizational constructs. Based on a new cyber threat USCYBERCOM directs all DOD organizations to do something that requires each cyber center to make a network configuration change.[33] USCYBERCOM issues an order which is modified and retransmitted by each Geographic and Functional Command, Service and Agency. The functional and service cyber centers receive the order from both the combatant command and their parent organization. Due to regional network configurations and varying technical and operational expertise in each of the staffs, the directions that reach the cyber center may differ significantly. At this point, the cyber center director must choose between joint and service direction. The cyber center director most often follows service direction since the joint chain of command neither provides funding, personnel, evaluation reports, nor sets day-to-day priorities. In the best case, the service implementation supports the joint direction. However, the service implementation may, in fact, work at cross purposes to what the joint commander is trying to achieve. The Government Accounting Office (GAO) highlighted this operational example in their recent report to congress concerning DOD cyber activities. The GAO cited "uncoordinated, conflicting and unsynchronized guidance ... leaving operators and administrators to reconcile priorities and ... procedures." [34]

The establishment of USCYBERCOM as the joint cyberspace planning, orders and synchronization authority has the potential to be a step in the right direction. However, the service cyber centers, the organizations which directly control the network, remain firmly established below service cyber commands. All funding, manning and day-to-day priorities are inextricably tied to the parent Service. The service cyber centers accomplish joint priorities in as much as the service priorities and service networks and systems are aligned with joint priorities. All joint force orders, whether they originate from USCYBERCOM, GCCs, or joint task force commanders, must filter through the Services before the service cyber centers execute them.

Despite the incredible efforts that the service cyber commands exert each day to operate and defend the network, their initiatives only marginally contribute to overall DOD cyber security and efficiency and support to the GCC.[35] While the DOD is pursuing the "seamless DOD Enterprise Information Environment (EIE)" that breaks down stove-pipe networks and duplicative command structures to support the joint warfighter, the Services are pursuing new and better service networks and organizations.[36] The Navy recently released a request for proposal (RFP) for the Navy's Next Generation Enterprise Network (NGEN) that will cost several billion dollars.[37] The Army also announced plans to create a "world-class cyberoperations center that will replicate cyberthreats."[38] While these initiatives may improve service cyber capabilities, they also create another network and organization that the GCC must deal with to integrate joint cyber operations.

In their article "Army, Navy, Air Force and Cyber – Is it Time For a Cyberwarfare Branch of Military?", Colonels Conti and Surdu argue that the DOD create a cyberwarfare service or branch of the military to develop and sustain dominance in the cyber domain. They state that the service cyber components are "ill-fitting appendages" in their Service due to their technical mission and the traditional service approach to leadership and management.[39] They also cite the lack of boundaries in cyberspace and vital need for effective centralized network control and cyber defense as reasons to consolidate cyber expertise in a single branch. The services' inability to retain and effectively employ cyber personnel reinforces their argument. The creation of USCYBERCOM, shortly after this article was published, provided the technical environment and employment opportunities that the authors advocate. However, their arguments for creating effective interfaces amongst the Services and the joint community and centralizing network operations and defense to execute cyber operations remain.[40]

Colonel David Hathaway proposes a less extreme approach to organizing cyber forces which provides more effective interfaces and centralized control. Based on interviews with combatant command J6s and USCYBERCOM leaders, Colonel Hathaway postulated that USCYBERCOM should adopt a hybrid U.S. Transportation Command (USTRANSCOM) and U.S. Special Operations Command (USSOCOM) mission command structure.[41] In his model regional cyber commands (RCC) exist in each the GCC AOR and are assigned to USCYBERCOM. The combatant command exercises tactical control (TACON) over the RCC in their AOR. The RCCs exercise operational control (OPCON) over the service networks and cyber centers in their AOR and serve as the primary liaison between the GCC and USCYBERCOM for cyber requirements.[42] This proposed command structure provides USCYBERCOM centralized control to act quickly and globally in cyberspace and decentralized planning and execution in support of the GCC. However, the Services maintain the responsibility for building and operating Service networks under the direction of USCYBERCOM globally and RCCs regionally.

Cyber Architecture

While the RCC creates a single focal point for cyber operations in the region, the disparate, subordinate service networks and cyber centers limit the ability to achieve unity of effort. Each cyber center and network reflects their parent service culture and network management approach. These service cultures are a large impediment to infrastructure consolidation and achieving a truly joint EIE at the operational level. When discussing the Army's migration to Defense Information Systems Agency (DISA) hosted email at the Army's 2011 LandWarNet conference, Brigadier General Kevin Nally, U.S. Marine Corps, director, command, control, communications and computers/CIO, and Lieutenant General William Lord, U.S. Air Force chief of Warfighting Integration and CIO, "insisted that their services, for reasons of pride and other considerations, would not stand for losing their 'usmc.mil' or 'us.af.mil' email addresses, at least not without some additional discussion."[43] The Army and the Navy, however, agreed that "efficiency and consistency" were more important than military email domain names.[44]

While the seemingly innocuous and non-mission critical email address debate highlights service cyber divergence, the subject of standardized architecture and equipment raises even more disagreement. The current technical and functional complexity of DOD cyberspace is a direct result of this disagreement. The DOD GIG, or the military network element of cyberspace, is comprised of more than 15,000 networks and over 7 million systems.[45] This environment developed incrementally over 30 years under service direction with no centralized planning or engineering.[46] The fact that the current network supports joint warfighting to the extent that it does is a testimony to the herculean efforts of operators, intelligence personnel, and communicators who cobbled networks together for mission accomplishment throughout the years. To improve support to the joint community and reduce network complexities, the DOD is establishing policies and programs that require the Services to consolidate infrastructure.[47]

Network executive agency is a possible solution to the lack of centralized architecture. Under this concept, the DOD designates Services as executive agents (EA) for combatant command headquarters in the DOD Directive 5100.03.[48] The EA dictates the network architecture and standards. During a panel discussion with the DOD and Service CIOs, Lieutenant General Lawrence, the Army CIO/G6, cited her experience of two Services in Italy building two sets of network infrastructure on one base while a third Service had plans to pay for and install a third.[49] She said that the Services need to, "Stop the madness!" and submit to EA direction.[50] While service EAs would create a single set of standards and network architecture within individual GCC AORs, the problem of a single, secure and interoperable GIG network architecture would remain. As Zimet and Barry observe in their overview of military cyberspace, "While all the Services recognize the GIG as the umbrella network under which they will operate, there is no commonality among them as to network architecture."[51]

The ongoing European Joint Enterprise Network, Joint Information Environment (JEN/JIE) project highlights the benefits and challenges of the EA concept. When Colonel Gerald Miller was the 2d Signal Brigade commander in charge of running the Army's European communications network he recognized that the Army was paying millions of dollars for the USEUCOM and U.S. Africa Command (USAFRICOM) headquarters IT and network contracts. These contracts duplicated based operating communications services that the Army was already providing to U.S. Army Europe (USAREUR) and other Army and joint organizations. The battalion that provided Army communications to the Stuttgart area was even located on the same kaserne as the USEUCOM headquarters. Because DOD assigned the Army as the EA for USEUCOM and USAFRICOM, Colonel Miller believed that the 2d Signal Brigade, and their parent Army organization, 5th Signal Command, should provide common user communications to both combatant command headquarters.

The JEN/JIE planners worked with the combatant command staffs, DISA and the National Security Agency and developed a common, secure architecture that could support the combatant command headquarters and continue to support USAREUR headquarters and Army units.[52] The technical solutions the planners developed clearly indicate that common network architecture is achievable.[53] However, the solutions only address the combatant command headquarters and Army networks and do not consider the other service networks in Europe. Also, the technical solutions did nothing to solve the most critical issues of mission command and supporting organizational structures.[54]

These unresolved issues highlight the main deficiency of the EA concept. The EA directive is simply administrative in nature and does not contain the command authority required to direct actions and enforce specific standards in cyberspace. The EA depends on a coalition of the willing and has no command or directive authority over other Services and Agencies in theater. The EA concept is effective for communications support which provides static capabilities such as network access, video teleconferencing, email and phone service. However, the dynamic nature of cyberspace demands a more centralized and authority based approach.

Another approach to the network architecture challenge is to appoint a single DOD command or agency as the chief architect. Under the current EA program, DISA is the chief architect. DISA is the EA for IT technology standards and responsible for "planning, engineering, acquiring, testing, fielding, and supporting global net-centric information and communications solutions" for the DOD.[55] In their EA role, DISA develops and publishes standards and conducts command cyber readiness inspections in order to verify compliance. The DISA standards documents and inspection criteria detail interoperability and security standards while still providing the Services quite a bit of implementation and operational latitude. Also, the fact that DISA is under the direction of the Secretary of Defense as a Combat Support Agency and not under USCYBERCOM removes them, somewhat, from the cyber chain of command.

Recommendations

Due to DOD's cyber dependence and recognized cyber operational risks, the DOD made some evolutionary changes. These changes include the creation of USCYBERCOM and service cyber commands. However, now that all of these commands are fully operational capable, the network operations and defense transformation within the Services and the joint community is progressing along a slow, evolutionary path. Little has changed at the combatant command and service network operations level to improve cyber operations and defense. However, a few revolutionary changes in this arena would make a huge impact toward achieving cyberspace unity of command and unity of effort. These revolutionary changes fall in the two broad categories of restructuring cyber forces and reengineering the network.

Restructure Cyber Forces. The current layered and duplicative cyber mission command structure inhibits USCYBERCOM's ability to conduct rapid and synchronized global cyber operations and defense and the GCC's ability to integrate cyber operations into their regional plans and operations. The geographic combatant commands, USSTRATCOM, combat support agencies and the military services all have cyber responsibilities. These duplicative and diffused responsibilities prevent effective, synchronized operations and defense at all levels. To reduce the mission command complexity, the DOD should implement Colonel Hathaway's recommended hybrid mission command structure. However, rather than the RCC directing regional Service cyber centers, the DOD should consolidate all regional and global cyber centers into joint cyber centers assigned to the RCC and USCYBERCOM, respectively. This consolidation creates a single, authoritative global cyber command with subordinate regional commands responsible for the five aspects of build, operate, defend, exploit, and attack in the cyber domain.[56]

In addition to establishing joint cyber centers, the DOD should remission the service cyber components. The new mission should be to man, train, and equip service cyber forces and provide tactical communications forces to service tactical units. The new service cyber missions would not include any responsibility for building, operating or defending global and regional networks. The service cyber responsibilities would mirror the current service special operations forces component responsibilities. Retaining the man, train and equip functions in the Services allows USCYBERCOM and the RCCs to capitalize on service cultures and experience while still providing Service specific tactical communications. Rather than forming a cyber service which is completely disconnected from the other Services as Conti and Surdu argued, this consolidation strategy capitalizes on the strengths of each service culture and ensures that the cyber personnel at USCYBERCOM and the RCC can identify with the warfighting requirements of the GCC and service components they support.

The DOD should also establish DISA as a subordinate command under USCYBERCOM. This move would unify cyber architecture, inspection and engineering authorities under a single chain of command. This reorganization would bolster the joint cyber center construct by consolidating the major build, operate and defend organizations from the joint service provider down to the regional operator. Also, the DISA command center could serve as the core for the USCYBERCOM global cyber center.

Reengineer the Network. The primary impediment to cyber center consolidation is an antiquated, service-centric approach to provisioning networks. Each Service builds, operates and defends duplicative global, regional, and post, camp and station networks. While the service and functional applications on the networks may differ, the fundamental network and communications technology are the same. There are no technical reasons that each Service must operate on separate networks.

Therefore, in conjunction with cyber center consolidation, the DOD must reengineer and integrate the service networks into a single DOD network per classification level.[57] The joint and regional cyber centers will build, operate and defend this new joint architecture under USCYBERCOM direction. This new architecture will move the DOD from a conglomeration of networks organized to form an enterprise information environment to a truly joint infrastructure. This integration also allows USCYBERCOM to conduct the fundamental reengineering that is required to implement an effective, interoperable and secure military cyberspace.

Implementation Requirements and Benefits. Consolidating cyber centers and networks into a single joint mission command construct and network will require centralized planning and phased execution. USCYBERCOM must direct the planning with each Service developing a subordinate consolidation plan. The central plan will detail the joint cyber center functions and manning requirements for building, operating and defending the network in support of the GCCs and regional forces and the GIG as a whole. This plan should capitalize on existing regional forces to form the core of the RCC and joint cyber center. For example, as Brigadier General Jeffrey G. Smith, Jr. advocated while he was the 5th Signal Command commander, 5th Signal Command and the 5th Signal Center, the Army regional cyber center, could form the core of the RCC and joint cyber center in Europe.[58]

The benefits of consolidating networks and cyber centers under the RCC and USCYBERCOM include streamlined cyber mission command, efficient employment of limited cyber forces and the ability to create an integrated cyber COP for the commanders at all levels. Consolidating command of the cyber centers and networks under USCYBERCOM and subordinate RCCs simplifies the currently inefficient cyber mission command structure and simplifies network administration. By administering all of the theater networks and commanding the people who operate them, the RCC commander can more effectively respond to regional GCC and component requirements while also maintaining accountability to USCYBERCOM for global responsibilities.

Consolidation also eliminates the multiplicative personnel requirements and reduces the strain on all of the Services to develop cyber warriors. The DOD can realize great efficiencies by reducing the four service global cyber centers and DISA's global cyber center into one. Most of the cyber functions that the service cyber centers perform can be combined to effectively perform the function and free some cyber personnel to fill the identified critical cyber shortages in the combatant commands.[59] For example, while each service cyber command is developing a cyber threat identification, analysis and defense capability, cyber threats do not vary so significantly by Service that they require a completely independent service cyber threat organization. Consolidating these cyber threat capabilities would achieve economies of scale and focus threat analysis and defense operations. While the USCYBERCOM cyber center may contain small cells dedicated to service specific functions, the efficiencies gained through consolidation outweigh the cost of maintaining separate service cyber centers.

Another efficiency gained under this construct is the ability to conduct global continuity of operations. Standardized regional cyber centers and architectures with a single parent command facilitate a regional cyber center passing control to USCYBERCOM or a sister region during planned maintenance operations or when they become non-operational due to unforeseen circumstances. Current cyber center diversity and incompatible architectures prevent such operations.

Finally, in addition to streamlining mission command and realizing efficiencies, network and cyber center consolidation enable the USCYBERCOM and RCC commanders to construct the, heretofore elusive, cyber COP. By removing the stove-pipe service networks and establishing a common architecture, USCYBERCOM will be able to construct an integrated view of the network, and, by combining functions such as the threat cell, detailed network threats and opportunities. The RCCs will be able to tailor the COP for regional requirements based on input from the GCC and component commands.

Risk

Restructuring cyber forces and reengineering the networks addresses several current risks while presenting some challenges. The following section of this paper analyzes these risks and challenges using the 2010 QDR risk assessment framework of operational, force management, institutional and future challenges risks.[60]

Operational Risk. Network and cyber center consolidation at the enterprise level directly addresses the DOD operational risk of securing systems in cyberspace.[61] Centralized mission command allows rapid and uniform action to counter threats at the global and regional level. This centralized control, however, can also pose an operational risk if it is not balanced by an understanding of GCC and component operations and requirements. Network consolidation reduces network complexity and enables more efficient administration. This homogenization, however, removes the current network redundancy and complexity which could potentially prevent a devastating attack.

Also, transition from the current service cyber center structure and network architectures poses an operational risk if not properly planned and executed. Cyber operations will be most at risk during the cyber center and network transitions. The DOD must judiciously execute organizational and network changes to minimize network service interruptions and resulting operational impacts.

Force Management Risk. Fundamentally altering the current service career paths also poses a risk to force recruitment, retention, training, education and equipping if not planned and executed properly. Given the shortage of cyber personnel, the DOD must consider the development and career progression of cyber professionals from all of the Services during consolidation. Career progression must include assignments with service tactical forces to capitalize on the strength of service cultures and operational experience.

Institutional Risk. Overcoming service cyber intransigence by removing the service requirement to operate and maintain the network presents enormous institutional risk. This approach requires a fundamentally different way of thinking about cyber space. The DOD must recognize the network as "a single, ubiquitous, centrally managed entity" rather than "a distributed network conjoined through diffuse pockets of geographic responsibility."[62] Many of the DOD business processes to include IT funding and organizational constructs must change. Overestimating costs and personnel requirements will doom this approach to irrelevance in the current resource constrained environment while underestimating either could produce disastrous operational consequences.

Future Challenges Risk. The ability to meet midterm and long term cyber threats and enable future mission success hinges on DOD's ability to transform its approach to cyberspace. Consolidating networks and cyber centers creates an effective, efficient and sustainable approach to meeting future cyber threats. Continuing to fund and develop service specific cyber infrastructures and organizations greatly increases the complexity of the problems and sustainment requirements.

Conclusion

Organizing and securing cyberspace as a warfighting domain is extremely challenging. This challenge is like building an airplane while in flight. Cyber defenders must counter the current deluge of attacks and cyber operators must enable the joint warfighter in accomplishing their missions. The DOD cannot simply turn cyber off, redesign and reequip it, and then turn it back on. Also, the Services cannot train a cadre of cyber experts overnight. Due to DOD's current cyber dependence, the transformation within the Services and the joint community, by necessity, is progressing along an evolutionary, rather than a revolutionary, path. However, a couple revolutionary changes would make a huge impact toward achieving cyberspace unity of command and unity of effort.

Consolidating networks and cyber centers under the RCC and USCYBERCOM greatly improves cyber mission command by significantly reducing the number of organizations in DOD that are responsible for building, operating and defending networks. This mission command structure establishes a single commander at the global and regional level who is responsible for the cyber domain. By reducing the number of organizations responsible for cyber and consolidating like functions across the Services, this concept allows for a more efficient employment of limited cyber forces. Finally, establishing a standardized network architecture and unified command and control structure enables USCYBERCOM and the RCCs to create an integrated cyber COP for the commanders at all levels.

The current threat environment coupled with DOD's dependence on cyberspace dictate major changes in current cyber organizations and architectures. With the impending budget cuts and current drive for efficiencies, the DOD cannot afford to fund separate infrastructures and duplicative organizations. As Lieutenant General Pollett, director, DISA, stated about the various military organizations that must gain efficiencies without compromising missions, "We no longer can afford to compete with each other."[63]

Endnotes

[1] Cyber operations, as defined by Army Cyber Command/2d Army, include the following functions: build, operate, defend, exploit and attack. (William "Max" Duggan, "Transforming Cyber While at War … We Can't Afford Not To," briefing presented at Defense Daily's 2011 Cyber Security Summit, Arlington, VA, Marriot Crystal City at Reagan International Airport, June 3, 2011.) This paper explores the build, operate and defend functions and does not consider cyber attack and exploitation. As such, "cyber" in this paper refers primarily to building, operating and defending network infrastructure and does not encompass all of cyberspace.

[2] The most significant documents are: 1. *The Comprehensive National Cybersecurity Initiative (CNCI),* (Washington, DC, The White House, March 2, 2010); 2. *National Cyber Incident Response Plan, Interim Version* (Washington, DC: U.S. Department of Homeland Security, September 2010); 3. *Enabling Distributed Security in Cyberspace: Building a Healthy and Resilient Cyber Ecosystem with Automated Collective Action* (Washington, DC: U.S. Department of Homeland Security, March 23, 2011); 4. Barack Obama, *National Strategy for Trusted Identities in Cyberspace: Enhancing Online Choice, Efficiency, Security and Privacy,* (Washington, DC: The White House, April 2011); 5. Barack Obama, *International Strategy for Cyberspace: Prosperity, Security, and Openness in a Networked World* (Washington, DC: The White House, May 2011); 6. *Cybersecurity, Innovation, and the Internet Economy,* (Washington, DC: U.S. Department of Commerce Internet Policy Task Force, June 2011); 7. *Department of Defense Strategy for Operating in Cyberspace* (Washington, DC: U.S. Department of Defense, July 14, 2011); 8. *Department of Defense (DOD) Information Technology (IT) Enterprise Strategy Roadmap* (Washington, DC: U.S. Department of Defense, September 6, 2011).

[3] Barack Obama, *National Security Strategy* (Washington, DC: The White House, May 2010), 27.

[4] M. G. Mullen, *The National Military Strategy of the United States of America* (Washington, DC: Office of the Chairman of the Joint Chiefs of Staff, 8 Feb 2011), 3. The national military objectives are: counter violent extremism, deter and defeat aggression, strengthen international and regional security, and shape the future force.

[5] Ibid., 10.

[6] Ibid., 9.

[7] Leon E. Panetta, *Sustaining U.S. Global Leadership: Priorities for 21st Century Defense* (Washington, DC: U.S. Department of Defense, January 5, 2012), 1.

[8] Ibid.

[9] Ibid., 3.

[10] Ibid., 4.

[11] Ibid., 5,8.

[12] Leon E. Panetta, *Department of Defense Strategy for Operating in Cyberspace* (Washington, DC: U.S. Department of Defense, July 14, 2011), 1.

[13] Ibid., 5.

[14] Ibid., 6.

[15] Ibid., 8-12.

[16] William J. Lynn, III, *Department of Defense (DOD) Information Technology (IT) Enterprise Strategy Roadmap* (Washington, DC: U.S. Department of Defense, September 6, 2011), 16.

[17] Ibid., v.

[18] Panetta, *Department of Defense Strategy for Operating in Cyberspace,* 1.

[19] James R. Clapper, "Remarks as delivered by James R. Clapper, Director of National Intelligence, Open Hearing on the Worldwide Threat Assessment, House Permanent Select Committee on Intelligence," February 10, 2011, http://www.dni.gov/testimonies/20110210_testimony_hpsci_clapper.pdf (accessed December 25, 2011), 3-4.

[20] Ibid., 4.

[21] Magnus Hjortdal. "China's Use of Cyber Warfare: Espionage Meets Strategic Deterrence," *Journal of Strategic Security* IV, no. 2 (2011): 1.

[22] Office of the National Counterintelligence Executive, "Foreign Spies Stealing U.S. Economic Secrets in Cyberspace, Report to Congress on Foreign Economic and Industrial Espionage, 2009-2011," October 2011, http://www.dni.gov/reports/20111103_report_fecie.pdf, (accessed December 14, 2011), i.

[23] Hjortdal, 8.

[24] Robert M. Gates, *Quadrennial Defense Review* (Washington, DC: U.S. Department of Defense, February 2010), 90.

[25] Ibid., 37.

[26] U.S. Department of Homeland Security, *Enabling Distributed Security in Cyberspace: Building a Healthy and Resilient Cyber Ecosystem with Automated Collective Action* (Washington, DC: U.S. Department of Homeland Security, March 23, 2011), 5.

[27] Panetta, *Department of Defense Strategy for Operating in Cyberspace*, 5.

[28] Peter J. Beim, "Network Operations: The Role of the Geographic Combatant Commands," in *Information as Power: An Anthology of Selected United States Army War College Student Papers, Volume 2,* ed. Jeffrey L. Groh, David J. Smith, Cynthia E. Ayers and William O. Waddell (Carlisle Barracks, PA: U.S Army War College, 2008), 127-129. After Lieutenant Colonel Beim published this article the Services changed the term "network operations" to "cyber operations." Other than the addition of a service cyber command at the tops of the operations organizational chart, the Services still adhere to the centralized and regional organizations depicted.

[29] U.S. Joint Chiefs of Staff, *Joint Communications System*, Joint Publication 6-0 (Washington, DC: U.S. Joint Chiefs of Staff, June 10, 2010), IV-6 – IV-7.

[30] The specific experiments and organizations that the combatant commands have developed are beyond the scope of this paper. U.S. European Command created a Joint Force Cyber Component Command

[31] USCYBERCOM Commander's Intent, https://www.cybercom.smil.mil/default.aspx (accessed October 19, 2011).

[32] U.S. Joint Chiefs of Staff, *Joint Communications System*, III-1.

[33] This example is simplistic and illustrative of technical instructions flowing through the current cyber construct. I avoided a specific change in order to not get entangled in technical details. However, this very situation occurred on a daily basis while I was the Corps NOSC director in Iraq during Operation Buckshot Yankee (joint name) or Operation Rampart Yankee (Army name).

[34] U.S Government Accountability Office, *DOD Faces Challenges In Its Cyber Activities* (Washington, DC: U.S. Government Accountability Office, July 2011), 7-8.

[35] These comments are in no way meant to disparage the hard work that the service cyber commands perform each day.

[36] Lynn, *DOD IT Enterprise Strategy Roadmap*, 1.

[37] Robert K. Ackerman, "NGEN Race Heats Up," *SIGNAL,* December 2011, 19.

[38] Rita Boland and Maryann Lawlar, "The Cyber Army of the Future," *SIGNAL,* October 2011, 70.

[39] Gregory Conti and John "Buck" Surdu, "Army, Navy, Airforce and Cyber – Is it Time for a Cyberwarfare Branch of Military," *IAnewsletter* 12, no. 1 (Spring 2009): 15.

[40] Conti and Surdu, "Army, Navy, Airforce and Cyber," 17.

[41] David C. Hathaway, "The Digital Kasserine Pass: The Battle Over Command and Control of DOD's Cyber Forces" July 15, 2011, http://www.brookings.edu/papers/2011/0715_cyber_forces_hathaway.aspx (accessed October 20, 2011), iv.

[42] Ibid., 18-19.

[43] Max Cacas, "DISA Seeks Ideas, Innovation, and Collaboration," *SIGNAL,* October 2011, 76.

[44] Ibid.

[45] Lynn, *DOD IT Enterprise Strategy Roadmap*, 2.

[46] Ibid., iv.

[47] Ibid., 1.

[48] U.S. Department of Defense, *Support of the Headquarters of Combatant and Subordinate Unified Command*, Department of Defense Directive 5100.03 (Washington, DC: U.S. Department of Defense, February 9, 2011). The definition of Executive Agency is, "The Head of a DOD Component to whom the Secretary of Defense or the Deputy Secretary of Defense has assigned specific responsibilities, functions, and authorities to provide defined levels of support for operational missions, or administrative or other designated activities that involve two or more DOD Components." This definition is from U.S. Department of Defense, *DOD Executive Agent*, Department of Defense Directive 5101.1 (Washington, DC: U.S Department of Defense, September 3, 2002, Incorporating Change 1, May 9, 2003).

[49] Cacas, "DISA Seeks Ideas, Innovation, and Collaboration," 76.

[50] Ibid.

[51] Elihu Zimet and Charles L. Barry, "Military Service Overiew," in *Cyberpower and National Security,* ed. Franklin D. Kramer, Stuart H. Starr, and Larry K. Wentz (Washington, DC: Potomac Books, Inc., 2009), 304.

[52] The Army planners included 2d Signal Brigade, 5h Signal Command, Army Regional Computer Emergency Response Team Europe (RCERT-E),Army CIO/G6, 9th Signal Command (Army)/US Army Network Enterprise Technology Command (NETCOM), and 2d Army/Army Cyber Command.

[53] This project only addressed the Non-secure Internet Protocol Router Network (NIPRNET) and Secure Internet Protocol Router Network (SIPRNET). The project did not address the many functional and coalition networks.

[54] My information on this program is dated with my departure from theater on July 7, 2011.

[55] U.S. Department of Defense, *Defense Information Systems Agency*, Department of Defense Directive 5105.19 (Washington, DC: U.S. Department of Defense, July 25, 2006), 2, 6.

[56] Hathaway, "Digital Kasserine Pass," 9. There may also be an opportunity to consolidate cyber intelligence functions globally and regionally in order to enhance cyber operations, more specifically the exploit and attack functions. However, the focus of this paper is on the build, operate and defend functions.

[57] The service and DISA networks should be consolidated into a single DOD network according to the classification level of the information processed on the network. I am not suggesting a single, multi-level security (MLS) network at this time. However, such a consolidation would be possible when the MLS network technology is developed.

[58] Jeffrey G. Smith, Jr., "Cyber Concept Briefing to U.S. Army Cyber Command, 2d Army," Wiesbaden, Germany, 5th Signal Command, April 20, 2011.

[59] U.S. Government Accountability Office, *DOD Faces Challenges In Its Cyber Activities*, 38.

[60] Gates, *Quadrennial Defense Review*, 90.

[61] Ibid. Securing DOD systems in cyberspace is listed as a key issue that poses risk to operational mission in the near term.

[62] Wes Andrues, "What US Cyber Command Must Do," *Joint Forces Quarterly* 59 (4th Quarter 2010): 117.

[63] Boland and Lawlar, "The Cyber Army of the Future," 71.